Country Things

Other books by Bob Artley

Memories of a Former Kid

Cartoons

Cartoons II

A Country School

A Book of Chores

Ginny: A Love Remembered

Country Christmas

Country Things

BOB ARTLEY

IOWA STATE UNIVERSITY PRESS / AMES

*This book is for my brothers, Dean and Dan,
who are also well acquainted with country things.*

© 1994 Bob Artley
All rights reserved

Authorization to photocopy items for internal or personal use, or the internal or personal use of specific clients, is granted by Iowa State University Press, provided that the base fee of $.10 per copy is paid directly to the Copyright Clearance Center, 27 Congress Street, Salem, MA 01970. For those organizations that have been granted a photocopy license by CCC, a separate system of payments has been arranged. The fee code for users of the Transactional Reporting Service is 0-8138-2650-0/94 $.10.

∞ Printed on acid-free paper in the United States of America

First edition, 1994

Library of Congress Cataloging-in-Publication Data

Artley, Bob
    Country things / Bob Artley.—1st ed.
       p.   cm.
    ISBN 0-8138-2650-0
    1. Farm equipment—Pictorial works. 2. Agricultural implements—Pictorial works. 3. Farm buildings—Pictorial works. 4. Farm life—Pictorial works. I. Title.
S676.A78   1994
631.3′0973′022—dc20   94-21239

Contents

ACKNOWLEDGMENT, *vi*

PREFACE, *vii*

1 The House, *2*

2 Around the Farmyard, *22*

3 The Barn and Its Surrounding, *36*

4 Tools and Implements, *52*

5 In the Fields, *70*

6 Preparation for Winter, *86*

7 Dealing with the Cold, *96*

8 Fun and Relaxation, *108*

9 Beyond the Farm Gate, *118*

ACKNOWLEDGMENT. The material in this book is selected from the cartoon series "Country Things" distributed by Extra Newspaper Features of Rochester, Minnesota, and appearing in publications in the United States and Canada.

Preface

This book can be regarded as a museum. It depicts a collection of artifacts from a time when farming was close to the soil. And just as the objects in a museum are collected from attics, sheds, and barns of old farms, so the drawings in this book were sketched from the actual objects, photographs, or in many cases from memory. Like collectors for a museum, I also haunted, and continue to haunt, farm museums with sketchbook and camera in hand.

Having lived with, observed, and in most cases used many of the items here illustrated, I feel very much a part of them. Due to this intimate knowledge, I could as I drew almost "taste" many of the things, being emotionally steeped in that time and filled with nostalgia for a lifestyle that for the most part is past.

It was a time when farming was defined as *husbandry* (my modern dictionary lists the word as "archaic or poetic")—when farmers, or *husbandmen*, were totally committed to and involved in the care of their livestock and soil. And, of course, women of the farm were an integral part of the total farming operation, not only in keeping the house and nurturing the family, but often doing their share in the fields and around the barns as well.

For those of you who have also had some experience in that past (our numbers are dwindling) when farming was a hands-on occupation involving the whole family, I hope that you too will be able to taste it as you peruse these pages.

And for you who are too young to identify with much of what you see here, and for you to whom this material is foreign to your experience, even exotic, I hope you will at least be able to find the contents of this book to be interesting historical notes of a time when farming was an involvement of the whole person—head, hands, and heart.

Artley Farm
Hampton, Iowa

Bob Artley

Photo by Jean Artley Szymanski

BOB ARTLEY has retired to the century farm that was his boyhood home and where he continues to draw, write, and paint. After studying art at Grinnell College and the University of Iowa, Artley's career as an editorial cartoonist began with the *Des Moines Tribune* and continued at the *Worthington* (Minnesota) *Daily Globe*. He has also worked in advertising and, with his wife Ginny, published two weekly papers. Artley is the author of several books, the most recent being a memoir, *Ginny: A Love Remembered.*

Country Things

1

The House

When stepping into an old farmhouse, back in the early part of the 1900s, one was apt to be met with a symphony of smells.

These odors, in varying degrees according to the season, were likely to be dominated by the scent of woodsmoke or burning corncobs from the kitchen range. There was the faint odor of kerosene, lye soap, or phenol (a disinfectant). There likely would be the fragrances of old wood, carpets, linoleum, rose petals drying in a vase on the plate rail, or a musty smell of old books.

Since the farmhouse of that period was a self-sufficient place, there might be the subtle smell of yeast from homemade bread, freshly churned butter from an oak barrel churn, applesauce, stewed tomatoes for canning, or dill pickles, jam, and jelly, to say nothing of the delightful blended aromas left over from a country dinner that maybe included a beef or pork roast or stewed chicken, along with perked coffee.

There might also be the clean smell of freshly laundered, starched, and ironed clothes, all folded and ready to be put away.

These are some of the sensual responses that the things depicted in this chapter evoke in me.

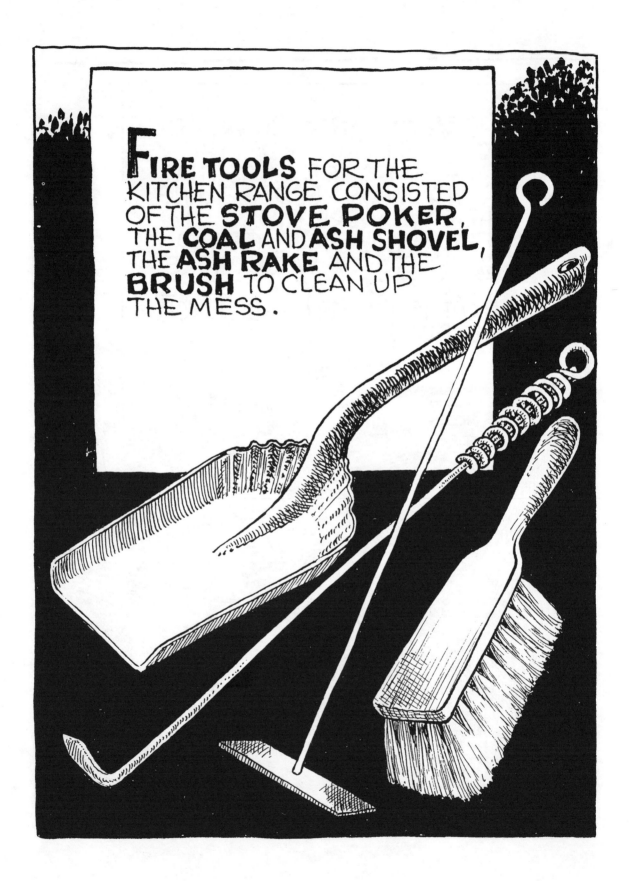

On the farm, with its full day of physical labor, we had **DINNER** at noon. Lunch was something to eat about mid-morning to tide us over until the noon meal, or in the mid-afternoon to keep us going until supper time.

There are few fragrances that warm the heart and excite the appetite, when coming in to supper from out of the cold, more than that of a juicy **PORK ROAST**, with potatoes and gravy, and a happy family with whom to share it.

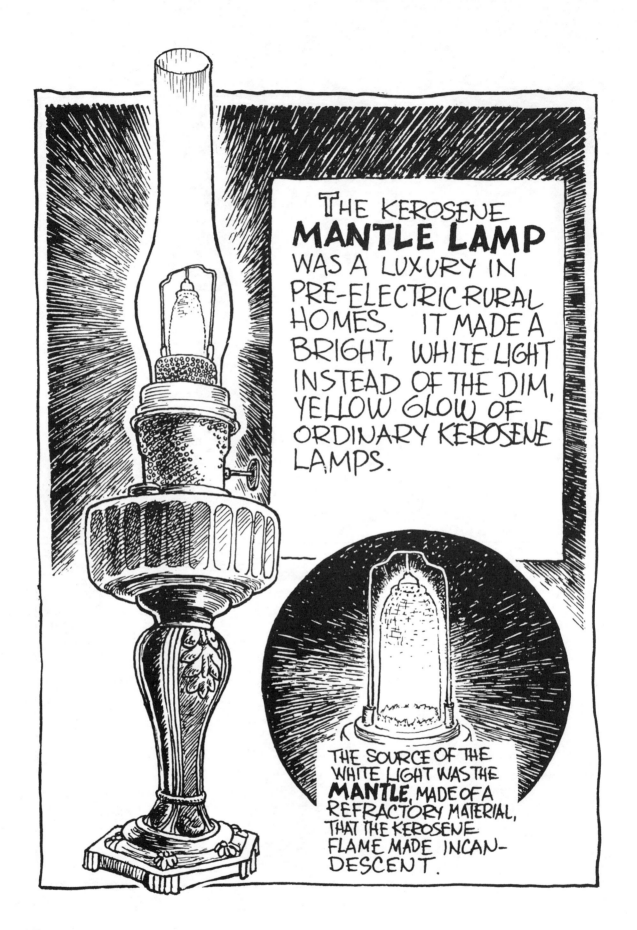

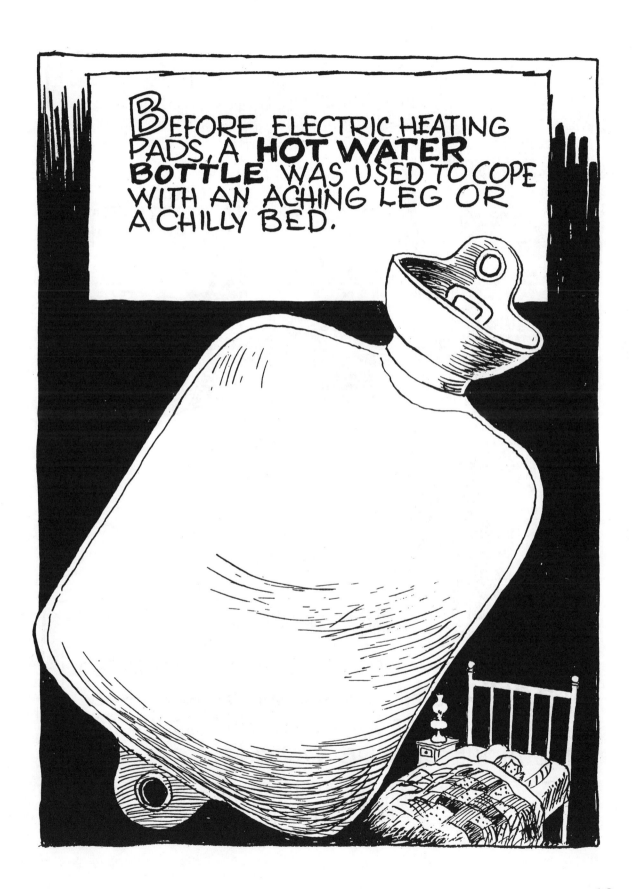

2

Around the Farmyard

The farmyard generally included the house yard and the barnyard—the area in which the farm buildings stood. It was the "campus" through which one moved from house to barn to granary to chicken house to hog pen, and so forth, when doing the daily farm chores.

It was in the farmyard that a visitor first arrived when coming up the driveway from the road. From that vantage point, one might observe much of the activity of the farm—there was always something going on.

One might see kittens playing in an open doorway of the barn, a rooster chasing a hen, sparrows flitting about, children playing, someone making garden, a mother hen scratching up edible tidbits for her covey of baby chicks, a cow drinking from the stock tank, someone repairing a wagon box or a barn door or a board fence, and from the pine trees one might hear the soft call of the turtledove.

Old apple orchards were often equipped with **CIDER MILLS** or **PRESSES**. Through theses mills fragrant, ripe apples were processed into sweet cider.

Properly stored the cider would stay sweet for some time. Otherwise it would ferment into hard cider and eventually vinegar.

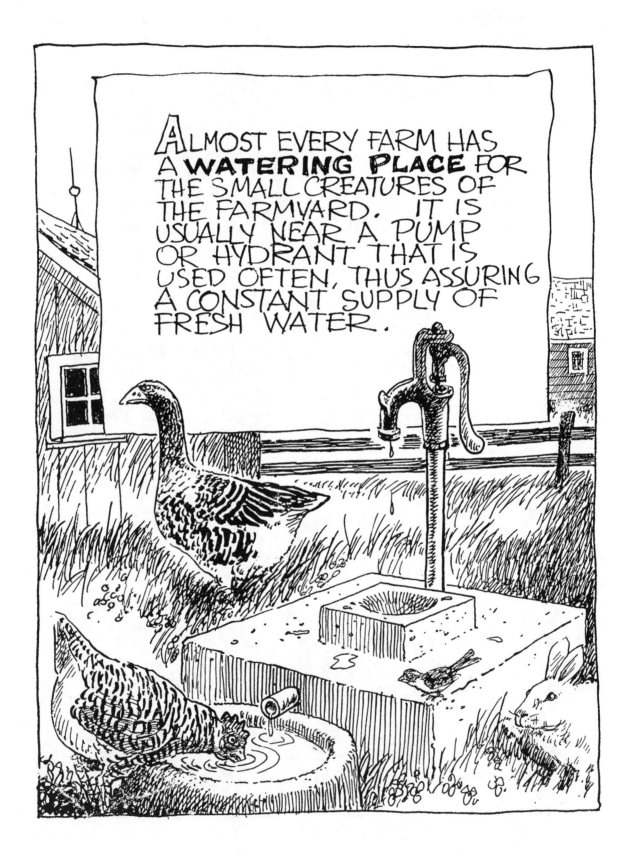

PROBABLY NOTHING SPEAKS MORE OF SPRING, TO THOSE WHO ARE FAMILIAR WITH OLD-TIME WAYS, THAN THE **SETTIN' HEN**. WHEN HER HORMONES CAUSED HER TO STOP LAYING EGGS AND START CLUCKING, SHE WAS PUT ON A NEST OF ABOUT 12 FERTILE EGGS. AFTER ABOUT 21 DAYS HER PATIENCE WAS REWARDED BY THE 12 EGGS TURNING INTO 12 CHICKS.

3

The Barn and Its Surroundings

The barn was easily my favorite building on the farm. There were many different kinds and sizes of barns to meet the various needs. However, the barn I have in mind is one that combined the storage of feed with living quarters for livestock, including horses, cows, sheep (pigs need a place apart), cats, pigeons, and, of course, whether we liked it or not—rats and mice.

To step into such a barn and inhale the fragrance of corn silage and hay mixed with the warm smell of horses, cattle, stored oats, and old timbers is a soul-satisfying experience for one who has grown up in such an environment.

The sounds, too, are music: the quiet shushing sound of milk being squirted into a nearly full, foamy pail, the creak of the stanchions as the cows lean for an elusive wisp of hay, the soft murmuring of the pigeons up in the loft, or the bawl of a young calf waiting a turn at its mother's udder for whatever is left after the cow has been milked.

All of this then is what I relive when I come upon an old milk stool, a scoop shovel, a hay rope pulley, or some artifact of that bygone era, or by some chance poke my head into a partly open door of an old barn, and catch a whiff that speaks of a time when life in that barn was active and full.

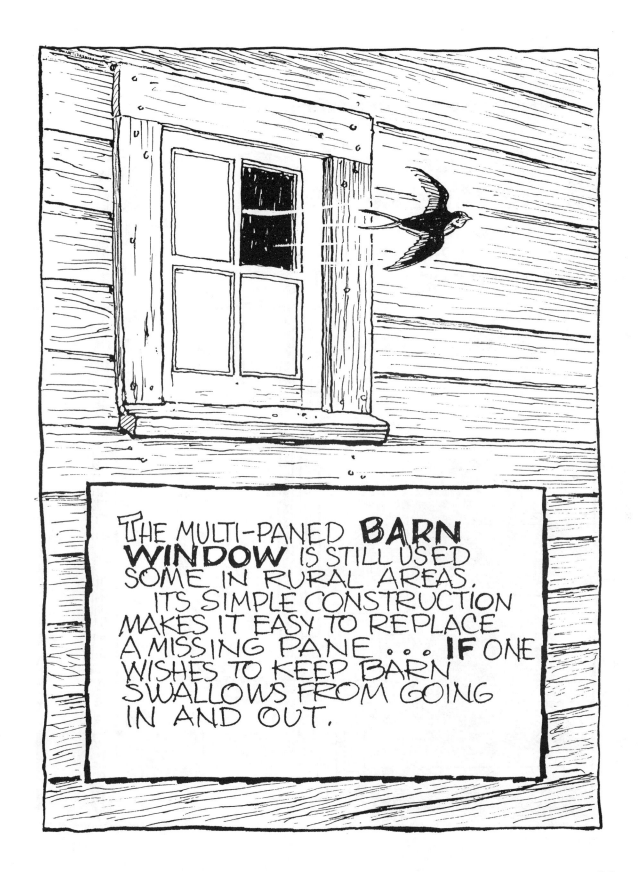

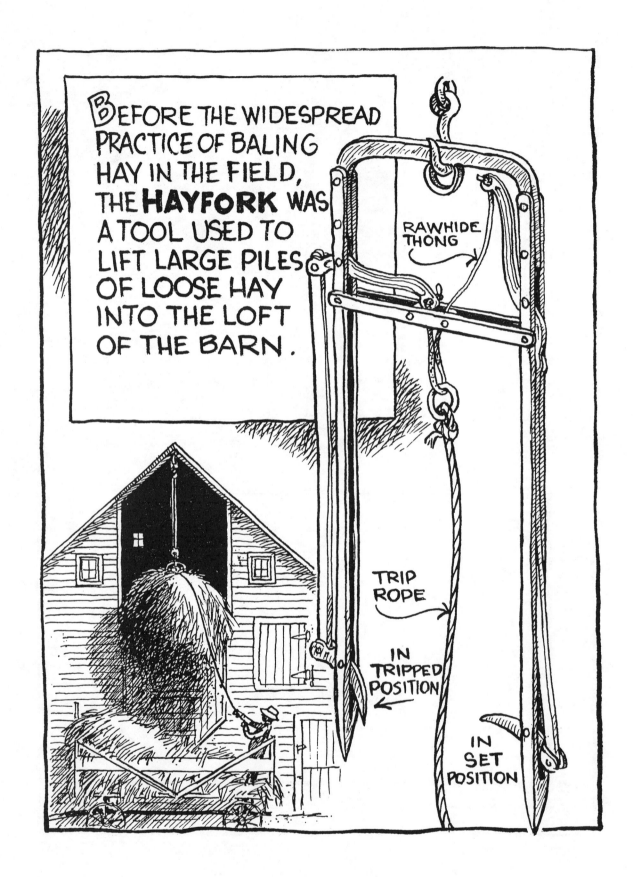

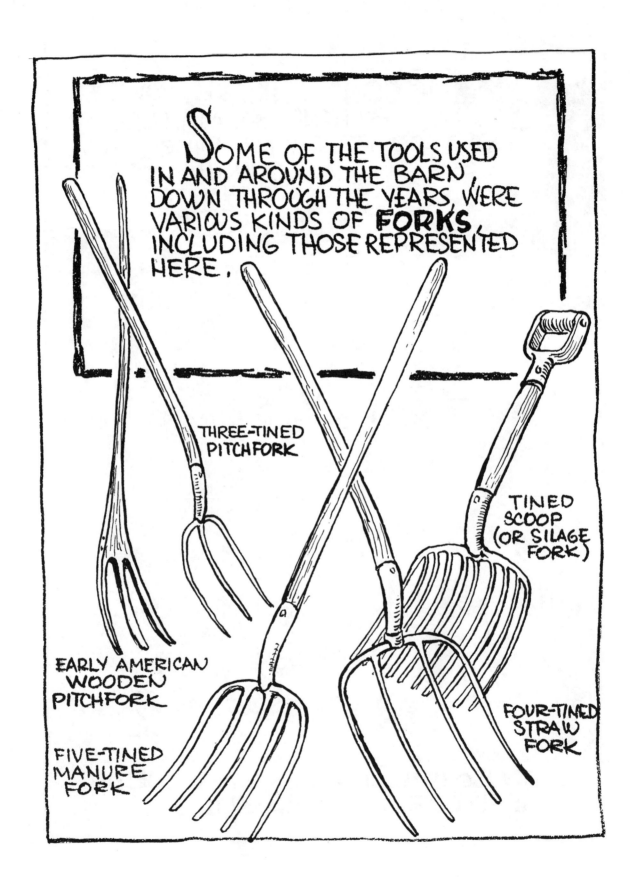

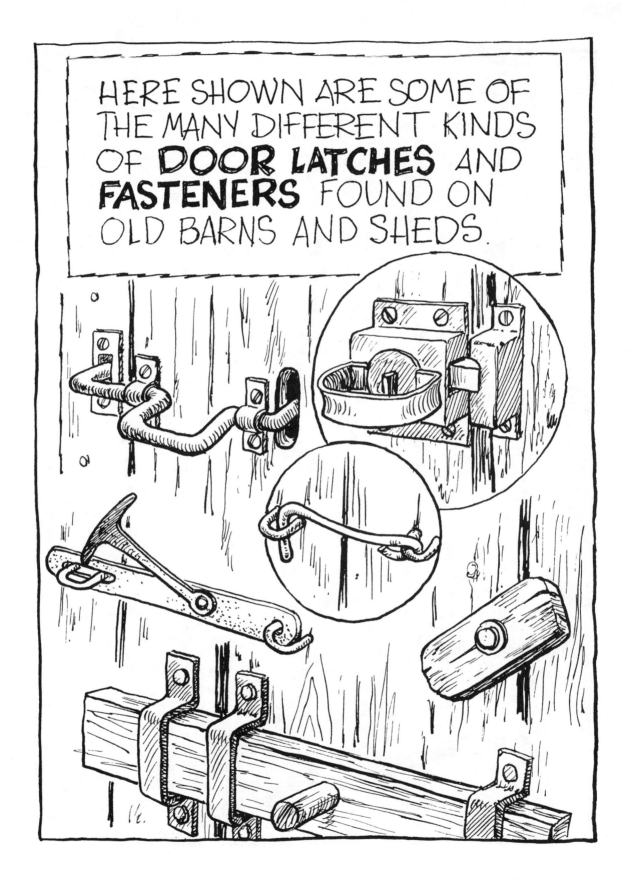

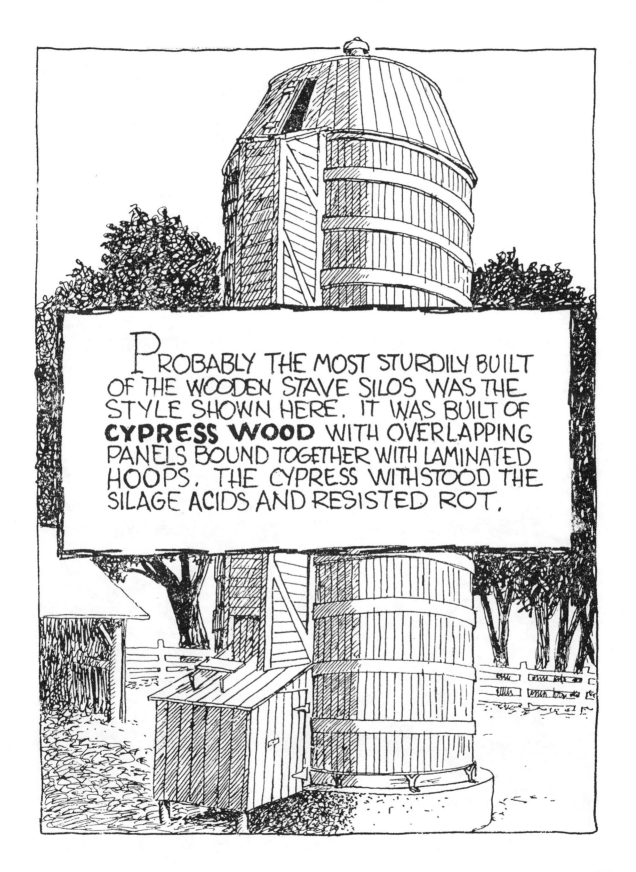

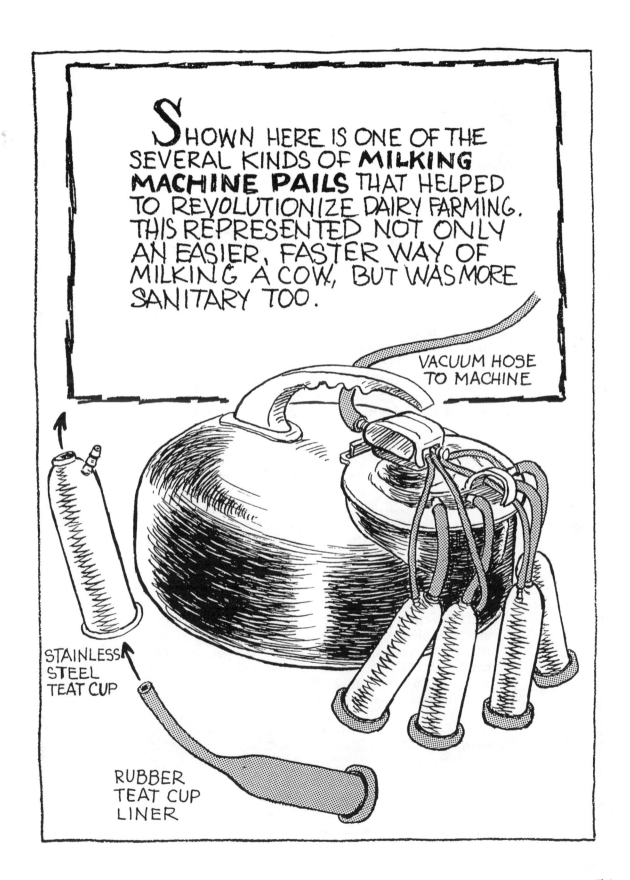

4

Tools and Implements

Next to the barn, the machine and tool shed was probably my favorite building on the farm. In summer its cavernous space was dark and cool and smelled of iron, wood, canvas, machine oil, and axle or cup grease. Kept there were most of the tools and equipment used around the farm: the tractor, feed grinder, oat seeder, corn planter, hay mower, oat binder, and corn binder—the machinery that was most vulnerable to weather.

For lack of space, machinery like the disc, drag harrow, and potato digger might be left outside at the mercy of the elements, the same as the hayrack with its steel running gears.

The wooden wagon box on its wooden chassis might find occasional shelter in the covered driveway between the corncribs or in the drive-through of the barn. The hand-powered corn sheller would be kept in the corncrib (where it was used), and the fanning mill in the granary or the feed room in the barn.

Tools, such as spades, shovels, hoes, rakes, axes, and crosscut logging saws, might also be kept in the machine shed or in a separate toolshed/workshop where the repairs were made. There, too, were the hand tools of the carpenter and mechanic, to be used when the farmer assumed those roles.

Pitchforks, manure forks, tined scoops, and scoop shovels were usually to be found in the areas where they were most used in the day-to-day business of farming.

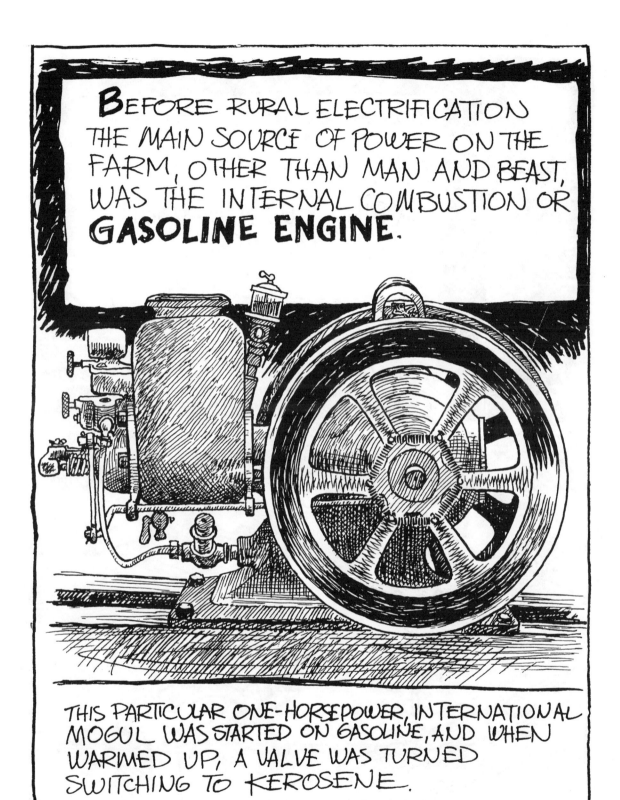

Before rural electrification the main source of power on the farm, other than man and beast, was the internal combustion or **GASOLINE ENGINE**.

This particular one-horsepower, International Mogul was started on gasoline, and when warmed up, a valve was turned switching to kerosene.

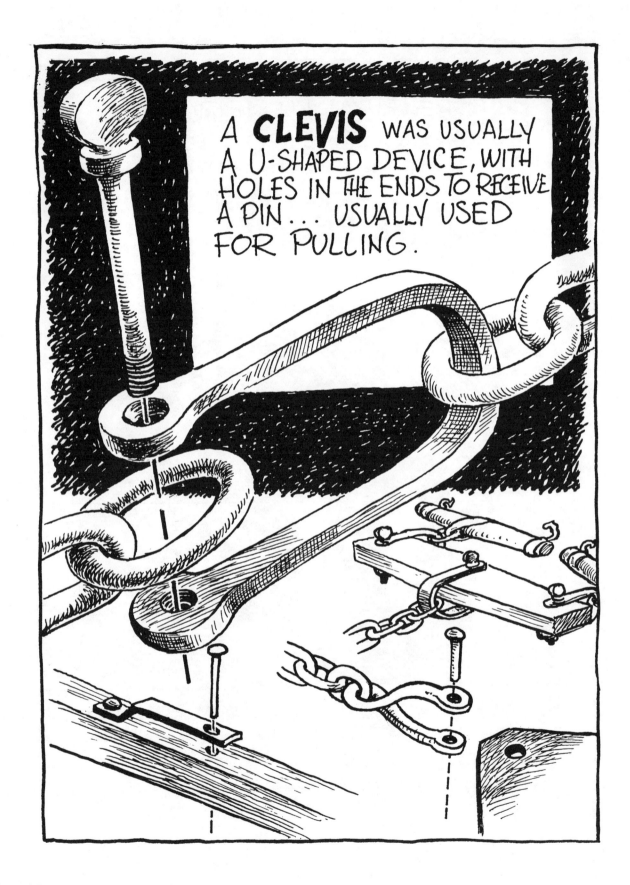

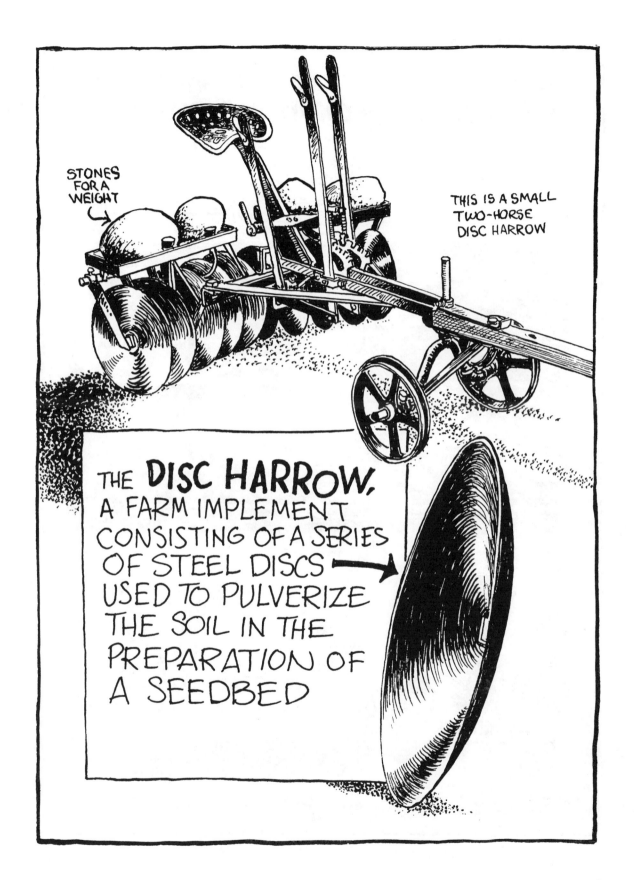

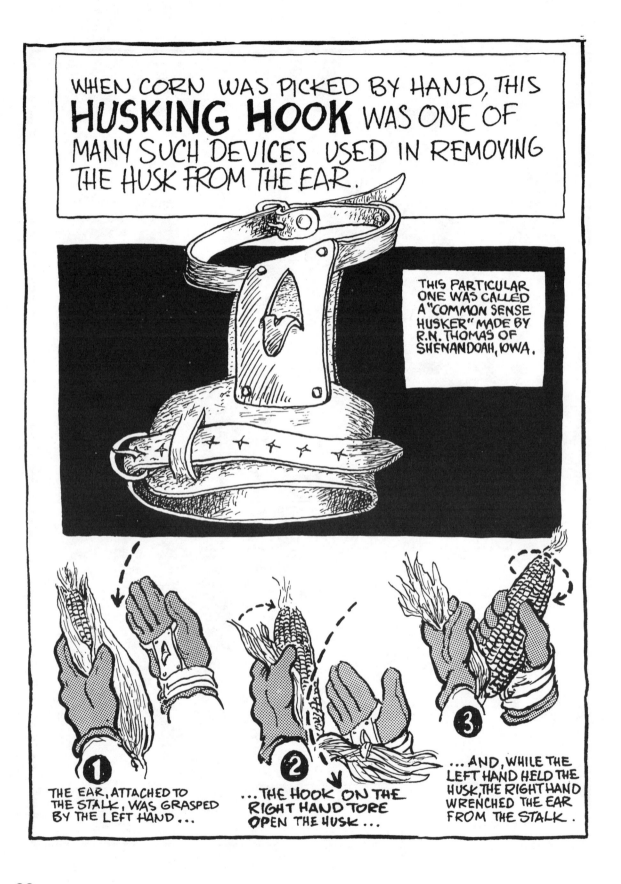

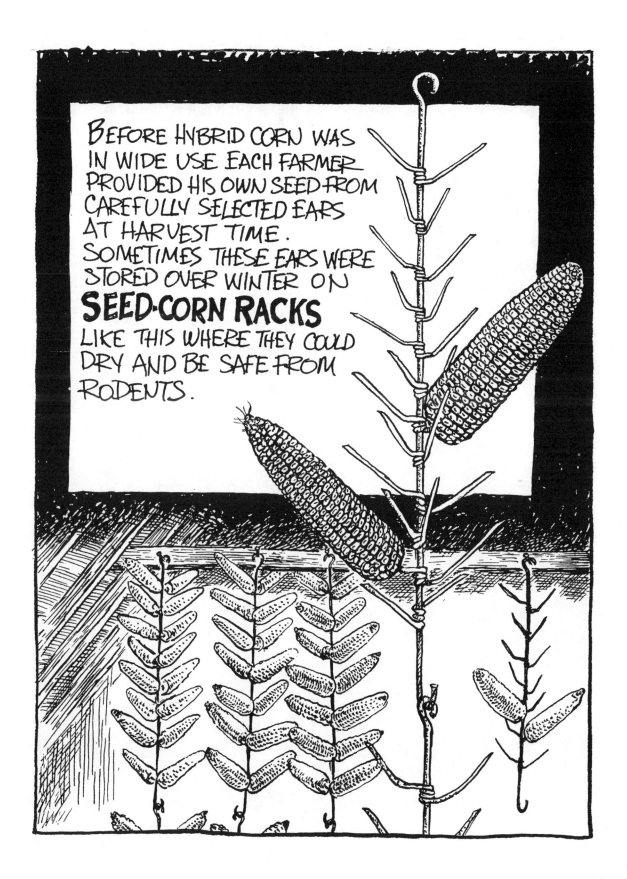

There was a time in the evolution of farming when the **THRESHING MACHINE**, or **SEPARATOR**, dominated the summer harvest scene. Here is shown a separator **FOLDED** into its **RESTING** or **QUIET** time.

But when the machine was in operation it was anything but quiet...

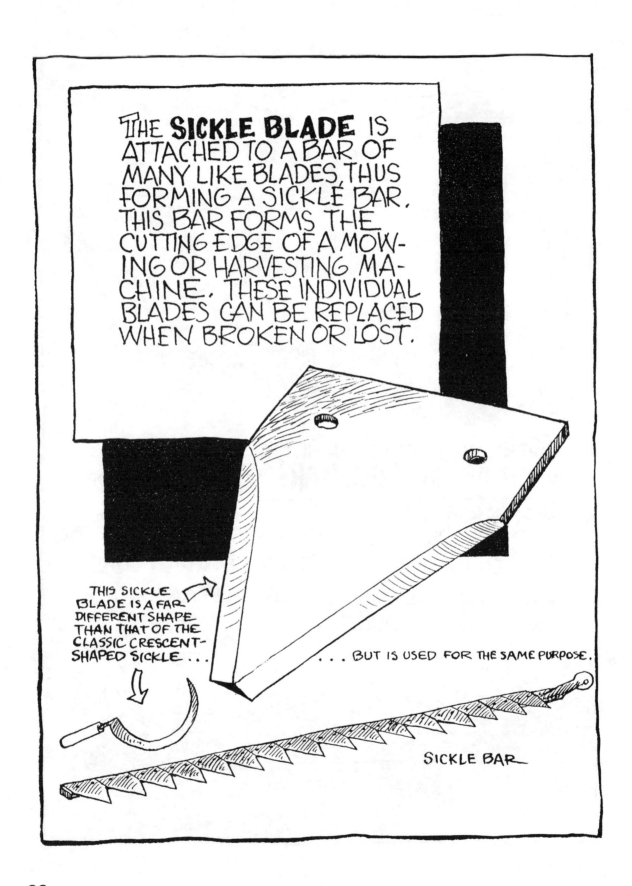

After the hay had cured in the swath (where it lay after being cut by the mowing machine), it was raked into windrows with the **DUMP RAKE** — and then raked into piles or cocks.

SWATH   WINDROW   HAYCOCK

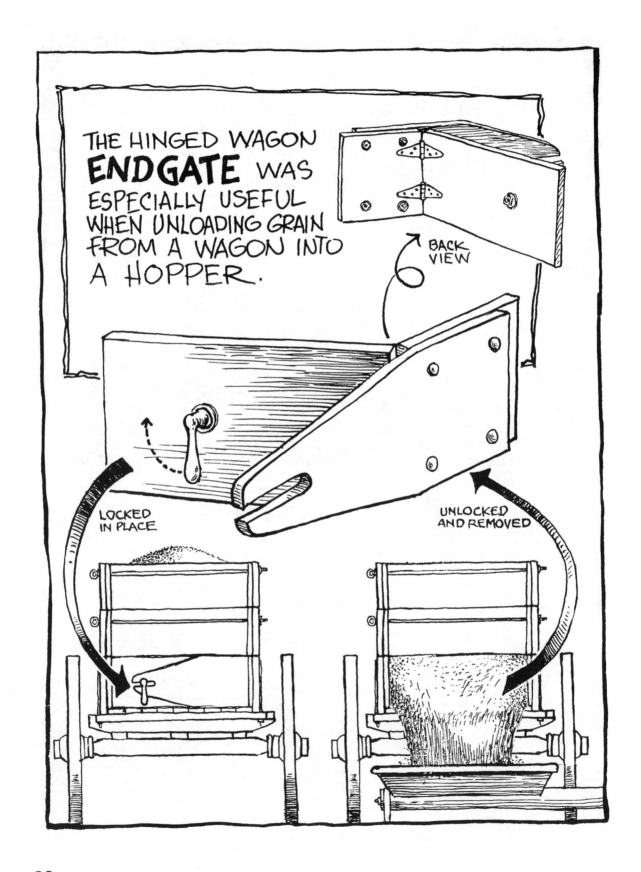

5

In the Fields

The phrase "in the fields" conjures up in my mind all kinds of pleasant images—images of openness under a bright sun with clean, fresh, sweet-smelling air carried on a gentle breeze that brushes through the grasses, wild flowers, and green corn leaves, the rollicking song of the meadow lark, the plaintive cry of the killdeer, and the graceful flight of the swallow skimming low over the heads of ripening oats as it gathers flying insects.

It also brings to mind muscle-straining work, sweat, a longing for the rest of noontime or evening, and the gnawing hunger for nourishing food that such labors bring on.

And, of course, I also see the approach of a violent thunderstorm as we scurry for shelter, and the harsh winter scene of wind-driven snow hissing through the protruding grasses and stubble of last season's crops.

No matter the season, in all of their variety the fields are a part of my longing when I dream of the farm that I knew when I was young.

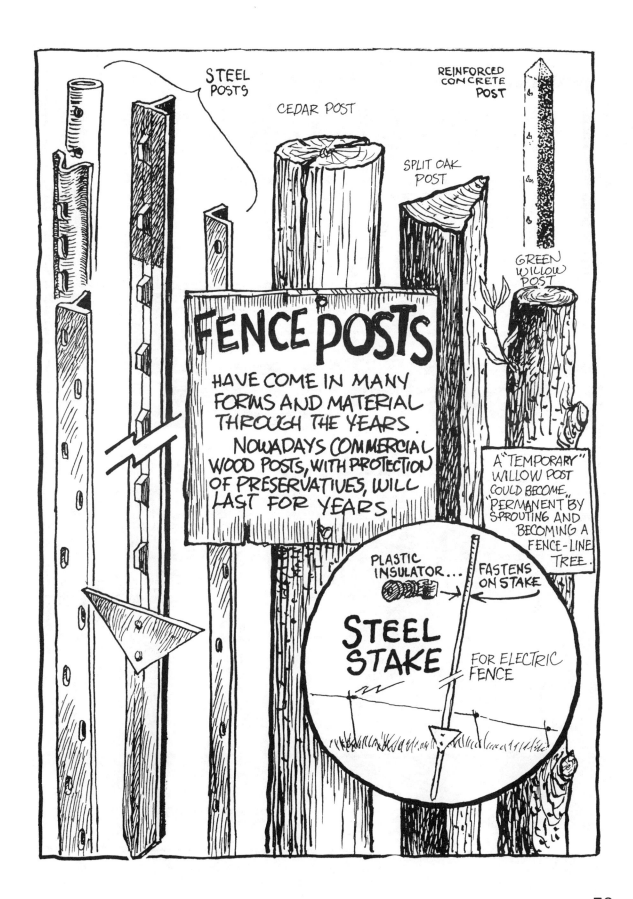

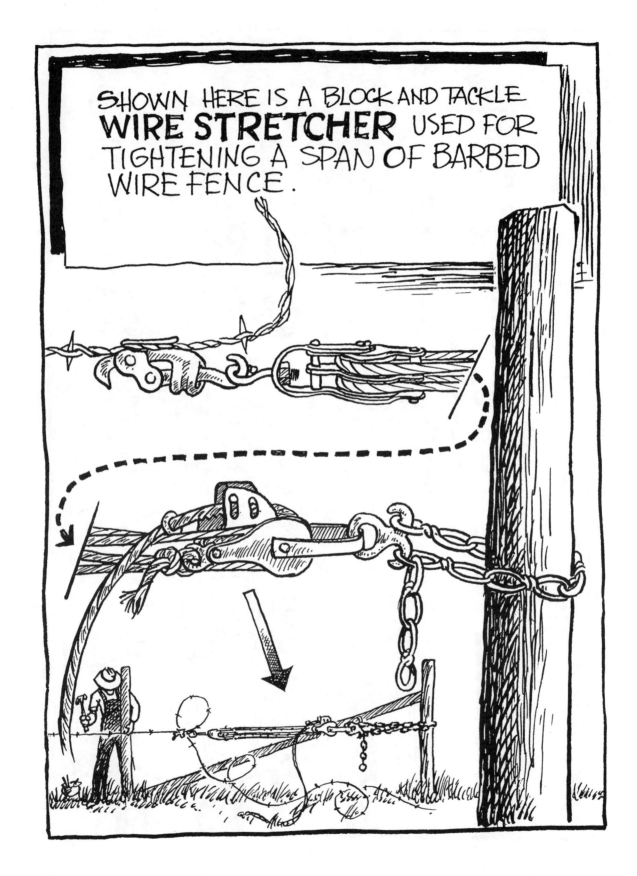

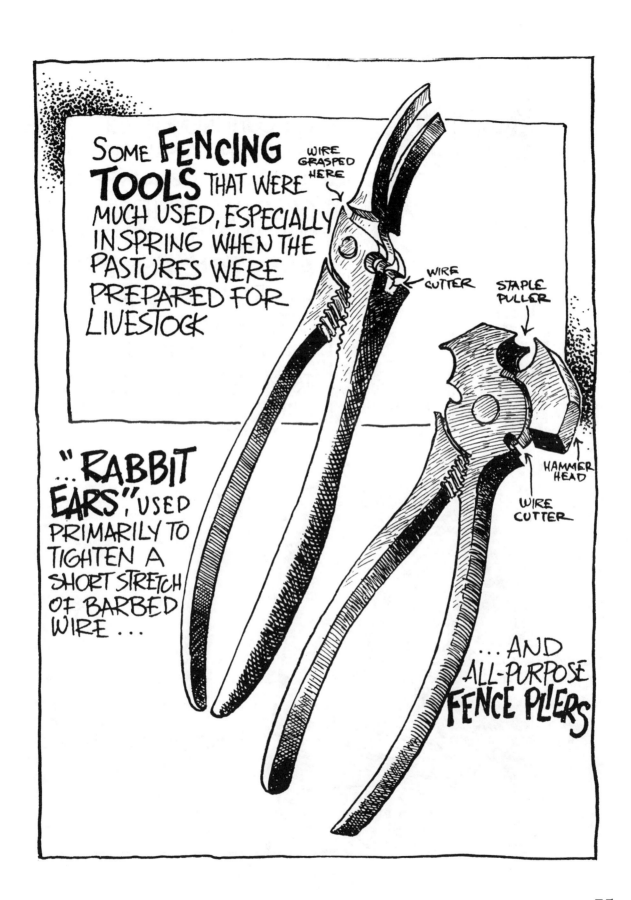

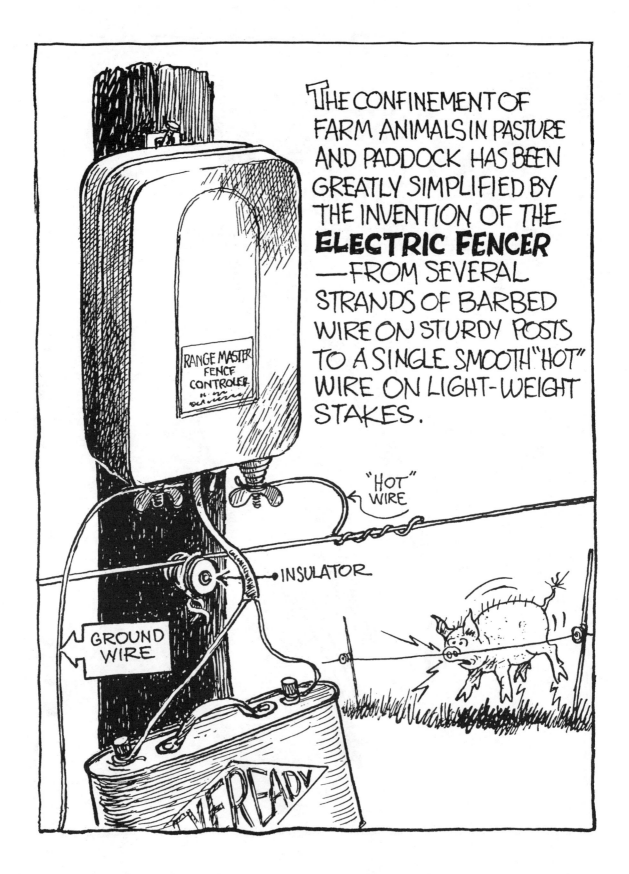

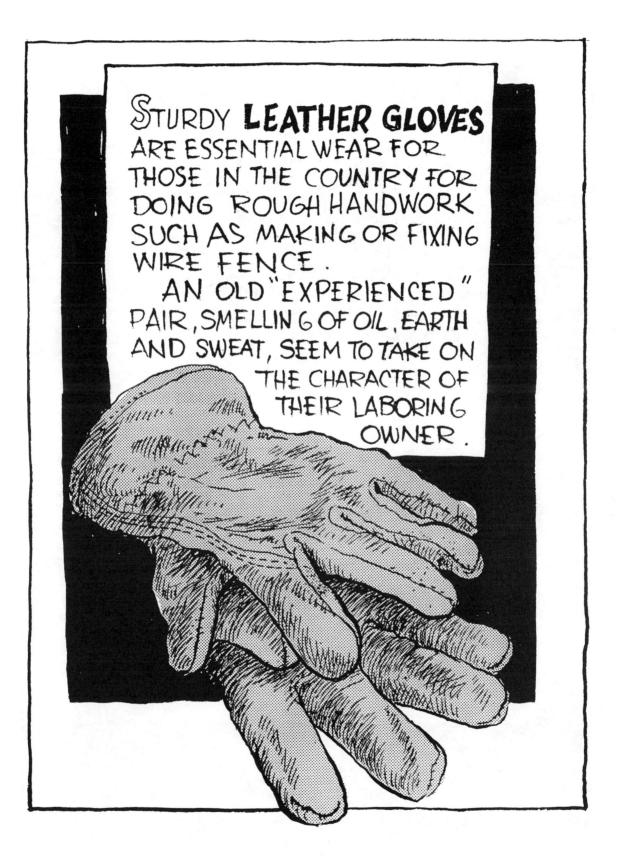

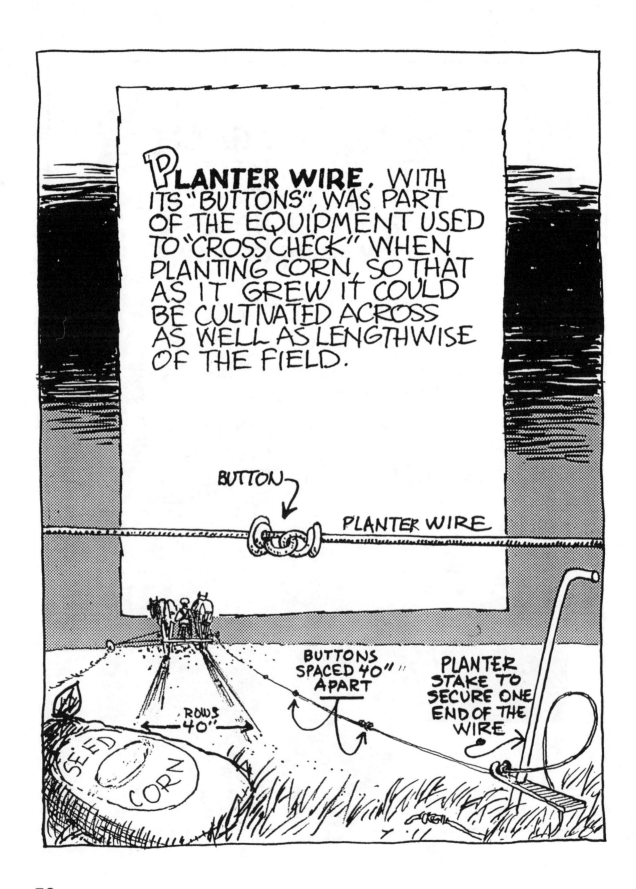

SUMMERTIME, THIRTY OR FORTY YEARS AGO, BEFORE THE WIDESPREAD USE OF THE COMBINE, THE **OATS SHOCK** WAS A COMMON SIGHT, DOTTING THE FIELDS BY THE HUNDREDS AS THE GRAIN CURED AND WAITED FOR THE THRESHING CREWS.

One way of keeping flies from biting the draft horses, as they worked in the fields, was the movement of the **FLY NETS** as the horses walked. These nets, made of leather or fiber cords, and sometimes with burlap, were put on the horses' backs over the harness.

This wasn't totally effective but it helped some.

6

Preparation for Winter

Autumn on the farm was a time of fulfillment and of urgency when the rewards of the spring and summer's labors were gathered and stored. If the crops were bountiful it was a time for being thankful.

There was a special ambience in the fields and orchards and kitchens in the fall, a quiet excitement in harvesting the fruits of the earth and storing them away against the oncoming winter—that time of death and dormancy when the land was locked in its icy grip.

Autumn was a time for making ready the "nest" as a place to survive the inhospitable winter months ahead. In years past the labors of autumn could be literally a matter of life or death. Perhaps it is this primeval urgency lingering in our collective memories that adds to the excitement of gathering and storing and making secure our nests.

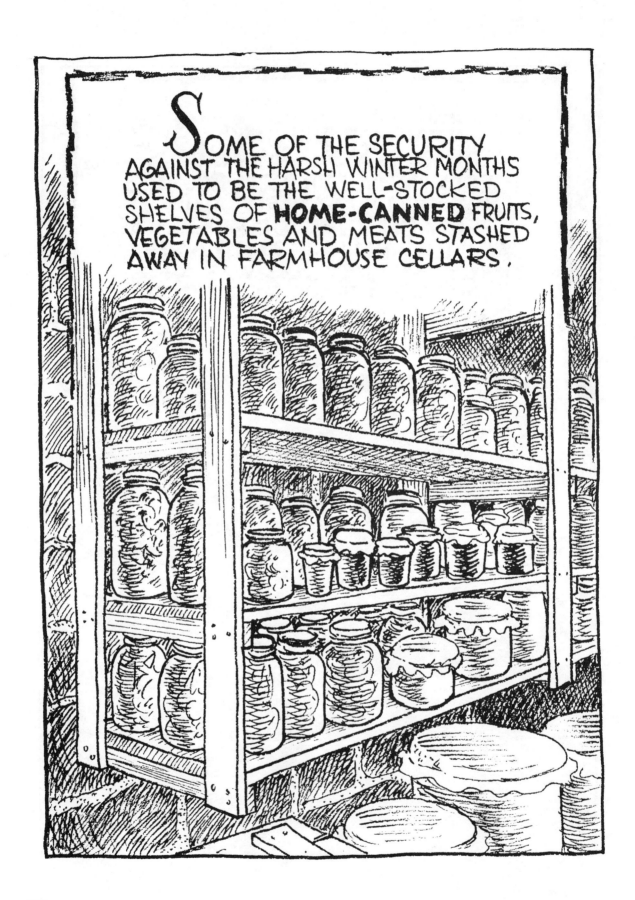

Our winter harvest of heating and cooking fuel included the task of **SAWING WOOD**. We cut the logs we had previously cut, split and gathered into piles, into stove-length chunks. The dry wood, for immediate use, was separated from that which had to season for about a year.

There was the same sense of accomplishment in **HAULING STOVE WOOD** in from the woodlot as there was bringing the summer harvest in from the fields.

7

Dealing with the Cold

From earliest times on this planet earth, humans have had to deal with weather in one form or another, searching out shelter in caves and wrapping themselves in animal skins or woven fibers.

More recently, but still a while ago—a time that most of the things in this book are about—there was a constant need to deal with the variety of temperatures and storm systems that swept through this so-called temperate zone in winter. This chapter is about some of the things that country people devised to cope with the cold. Whether it was stacking stove wood for our winter fires or making snug and dry the barns and stalls for the livestock, I felt a warm satisfaction in my labors to make things cozy for all of us creatures on the farm.

In the 1920s and 1930s the **SCOTCH CAP**, made of closely woven, knitted wool cloth, was popular with farmers as warm winter wear.

The inside band could be turned down to cover the ears and back of neck.

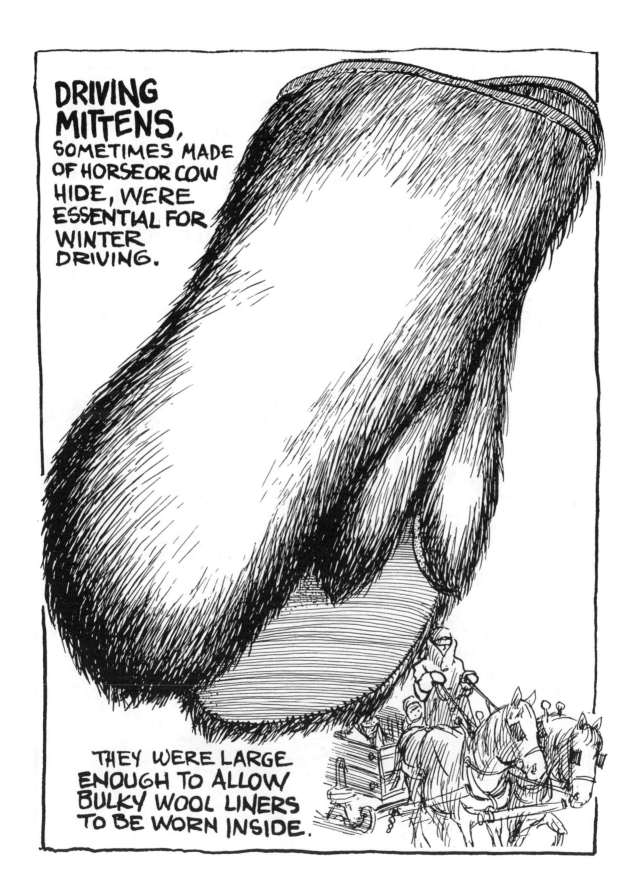

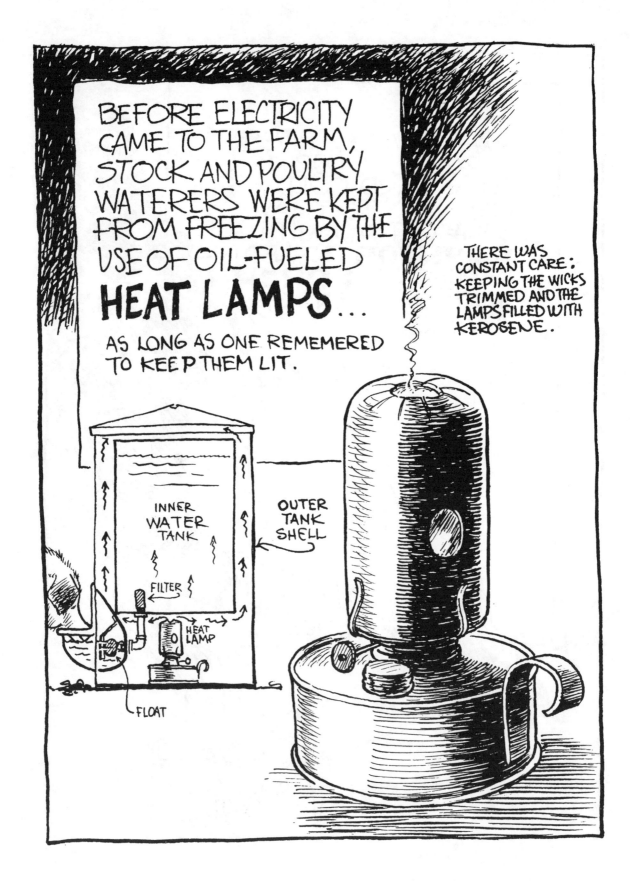

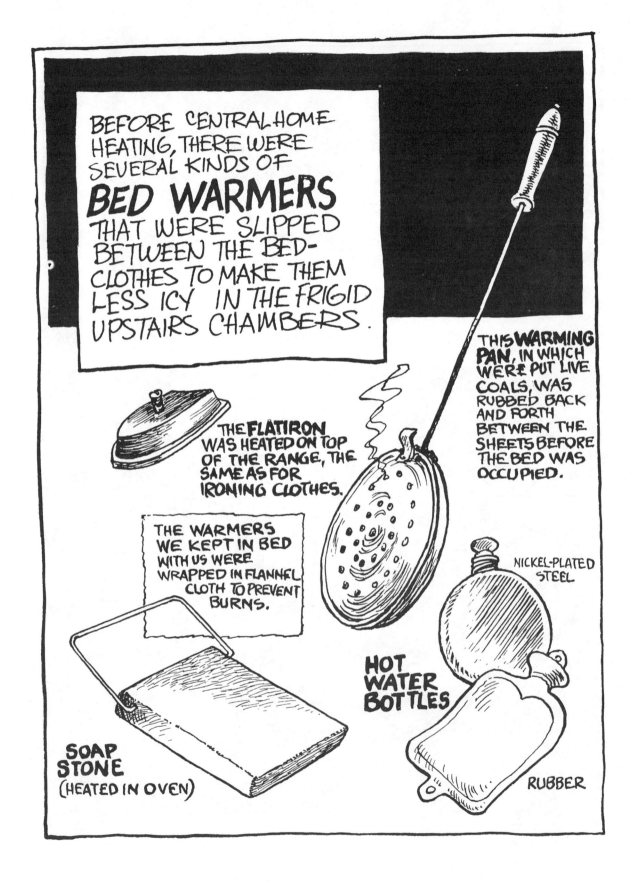

8

Fun and Relaxation

With a lively imagination, one can find things of interest and fun almost anywhere around the farm. This is especially true of children . . . and of adults who are fortunate enough to have a childlike spirit of wonder and adventure.

During the early decades of the 1900s, country children were mostly on their own when it came to finding fun things to do. With helpful, sympathetic adults nearby, those years could be a time for building rich memories to draw upon and nourish the soul in later years. We kids found great enjoyment simply in exploring all of the interesting, out-of-the-way places around the farm and looking for any adventure we might find there.

The following pages illustrate a few of the things that were employed to make those years a pleasant time to remember.

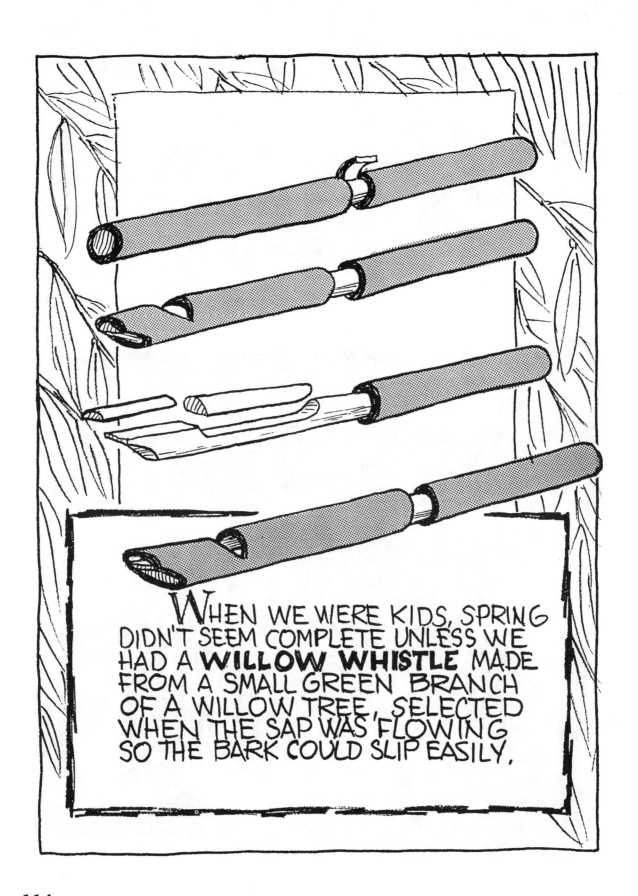

When we were kids, spring didn't seem complete unless we had a **WILLOW WHISTLE** made from a small green branch of a willow tree, selected when the sap was flowing so the bark could slip easily.

9

Beyond the Farm Gate

During my first four or five years, the house and farmstead were about all I experienced of the world. We had no car during those years, and when I did venture beyond those familiar surroundings it was with my parents in horse and buggy on muddy roads, or by team-drawn bobsled through a snow-filled countryside.

When I was about six or seven, my parents bought a Model T Ford coupe and my universe began to expand. About that time, I also started attending a country school about two miles from our farm, and a couple years later when my brother started school the two of us became explorers, not only of our own fields and streams and meadows but of the surrounding neighborhood as well. We found these adventures exciting, giving us a taste for exploring even farther from home.

In this chapter are some of the things that became familiar to us as our horizons expanded beyond the farm gate.

We weren't sure whether the name **SPRING WAGON*** was one that referred to the leaf springs that made for an easier ride over the rutted country roads, or to the fact that this horse-drawn vehicle was the surest way of negotiating the muddy roads of spring... and this was as recently as the first two decades of this century.

* TODAY WE MIGHT HAVE CALLED IT A PICKUP.

Spring often brought wet, chilly weather, sore throats, fevers, measles, whooping cough and all manner of ills. This, in turn, brought the family doctor down muddy roads into soggy farmsteads to dispense his pills and potions... and hope.